A BRIEF HISTORY OF ROADS IN VIRGINIA

1607-1840

Virginia Genealogical Society
Richmond, Virginia

Published with Permission from the

Virginia Transportation Research Council
(A Cooperative Organization Sponsored Jointly by
the Virginia Department of Transportation
and the University of Virginia)

HERITAGE BOOKS
2011

HERITAGE BOOKS
AN IMPRINT OF HERITAGE BOOKS, INC.

Books, CDs, and more—Worldwide

For our listing of thousands of titles see our website
at
www.HeritageBooks.com

Published 2011 by
HERITAGE BOOKS, INC.
Publishing Division
100 Railroad Ave. #104
Westminster, Maryland 21157

International Standard Book Numbers
Paperbound: 978-0-7884-3674-1
Clothbound: 978-0-7884-8897-9

A BRIEF HISTORY OF THE ROADS OF VIRGINIA 1607-1840

By

Nathaniel Mason Pawlett
Faculty Research Historian

Virginia Highway & Transportation Research Council
(A Cooperative Organization Sponsored Jointly by the Virginia
Department of Transportation and the University of Virginia)

Charlottesville, Virginia

October 1977

Revised November 2003

VHTRC 78-R16

The Historic Roads of Virginia Series

Louisa County Road Orders 1742-1748, by Nathaniel Mason Pawlett

Goochland County Road Orders 1728-1744, by Nathaniel Mason Pawlett

Albemarle County Road Orders 1744-1748, by Nathaniel Mason Pawlett

The Route of the Three Notch'd Road, by Nathaniel Mason Pawlett and
    Howard Newlon, Jr.

An Index to Roads in the Albemarle County Surveyors Books 1744-1853, by Nathaniel
    Mason Pawlett

A Brief History of the Staunton and James River Turnpike, by Douglas Young

Albemarle County Road Orders 1783-1816, by Nathaniel Mason Pawlett

A Brief History of the Roads of Virginia 1607-1840, by Nathaniel Mason Pawlett

Library of Congress Catalogue Card
No: 77-87261

PREFACE

As previously outlined in the publications of the series "Historic Roads of Virginia", this study grew out of a program of research into the history of road and bridge-building technology in Virginia undertaken by the Virginia Highway & Transportation Research Council in late 1972. This research has produced, besides this series, another called "Metal Truss Bridges in Virginia 1865-1932" by Dan Grove Deibler.

The purposes of the road history portion of this project are two: (1) to produce a history of the development of the roads of Albemarle County, Virginia, from the beginning of settlement about 1725 to the beginning of the turnpike era with the creation of the Board of Public Works in 1816; and (2) more importantly, to use this experience to develop a guidebook [and an advisory program] to assist local historical groups and interested individuals in doing the same thing for the other counties of Virginia.

This volume, "A Brief History of the Roads of Virginia 1607-1840", consists of what was originally intended. as the introductory chapters of the Albemarle road history. Since most readers are probably unfamiliar with the history of roads in Virginia, it was thought proper to devote the first section of that work to a sketch of the development of road transportation here up to the coming of the period of intense railroad development in the nineteenth century. At the suggestion of several of the people who read the first draft, and in-the interest of increasing the utility of this particular section, a separate publication was decided upon.

The author hopes that this brief sketch will help .to place development of the roads of Albemarle, as well as those of the other counties, within the larger context of the development of Virginia's roads, and that it will simultaneously provide some understanding of the varied and often conflicting forces which shaped transportation policy at the colonial and state levels.

"The past is a foreign country: they do things differently there",

L. P. Hartley, <u>The Go-Between</u>

THE COLONIAL PERIOD
1607-1776

THE COLONIAL PERIOD
1607-1776

Edward Graham Roberts provides the key to understanding the development of the roads of Virginia in a statement in his 1950 doctoral dissertation at the University of Virginia:

> The colonial Virginian was an Englishman. All his traditions were English. Until his evolution into an American was accomplished, his natural instinct was the recurring use of English ways and manners. Accustomed practices were followed until the impact of the wilderness created a native American whose problems were indigenous and demanded original solutions. Public convenience and welfare in the matter of transportation had to be met soon after settlement left the river banks. The instinctive recourse was to the Common Law and the basic road law of 1555 which had placed the responsibility for the construction and maintenance of roads upon the Parish, the smallest English administrative unit.

The 1555 act required six days' labor a year from each parishioner to help maintain the roads. Readers of newspapers may recall that the late Dame of Sark still possessed this power over the men on the Isle of Sark in the English Channel when she died.

This was the system which was brought to Virginia and established here. Throughout the colonial period the responsibility for the construction and maintenance of roads remained at the local level, first with the parish and later with the gentlemen justices of the county courts, though the ultimate responsibility rested with the individual surveyors, or overseers, of roads.

If the incidence of statutory enactments is any indication, roads received little attention during the first quarter century of Virginian history. By the end of that period, however, the "starving time" was past, a workable government had evolved and with it the likelihood that the Indians would not be able to turn back the tide of white settlement. By the end of this period settlement had begun away from the banks of those magnificent waterways which were to be the mainstay of tidewater transportation until about 1700. With the new settlements came the necessity for roads to connect them to water.

The first law dealing with roads in Virginia, also the first such law in America, appeared on the books in 1632. This law, following English precedent, placed the responsibility for roads on the parish. In England this system was probably rather well-suited to the needs of the times, for there the parish was a small administrative unit under the Anglican Church possessing few roads and a more than sufficient number of people to maintain them. Unfortunately, in Virginia parishes tended to be of immense size with sparse populations scattered over them. To clear and open, much less maintain, the network of roads required as settlement progressed was to prove beyond the capabilities of the existing system of administration.

Consequently, after 1657 road operations were placed in the hands of the gentlemen justices of the county courts, who appointed individual overseers of roads to handle specific roads or portions thereof. Further enactments in 1661 and 1663 reinforced that of 1657 directing that "convenient wayes" be constructed leading to parish churches, the county courts, to Jamestown (then the seat of government) and between the individual counties. Actual construction and maintenance were to be placed under the supervision of "overseers of roads" or "surveyors of highways", who were to be appointed annually by the county courts. All titheables, males above the age of 16 whether free or slave, were to work on the roads under their supervision when so ordered by their parish vestry. The vestry usually took this action upon a request being made by the individual surveyor. To forestall neglect of these duties, a schedule of fines was set out in the act. This schedule covered all the persons having to do with road work from the justices of the peace upon the bench, through the overseers of roads down to the "labouring male titheables", as they were styled, who actually did the work.

That these laws were something more than mere expressions of sentiment by the Burgesses is shown by the fact that such a man as Colonel Charles' Hill of Charles City County was fined for neglect of duty in 1676. In 1690 Governor Francis Nicholson issued a directive to the county courts ordering the justices to see that road surveyors were appointed and that they per- formed their duties or faced the penalties prescribed by law. Even a casual perusal of the order books of a Tidewater or Piedmont Virginia county in the eighteenth century will show, by the number of entries relating to indictments and prosecutions, that roads were a subject of continual interest for the gentlemen justices in the counties. Sometimes the county court itself was presented by the grand jury for failure to discharge its responsibilities relative to roads..

To a modern observer this system would appear exceedingly inefficient and the roads it produced little more than cleared tracks through the forests, but these roads were probably adequate to the needs of seventeenth century Virginians and the system was all the colony's resources would have allowed at the time. Wagon roads had not yet become the necessity that they would when settlement pushed past the fall line on the rivers. Wealth had not yet accumulated to the point where riding chairs, chariots and coaches could be afforded by the Virginian planters. Therefore a path suitable for travel on horseback usually sufficed for most Virginians. What few wagon roads and cartways there were served as transportation links between the inland plantations and water communication with the capitol at Jamestown, as well as such English ports as London and Bristol. This east-west orientation persisted throughout the eighteenth century; it was only with independence in 1776 that the transportation network gradually began to reorient itself from an east-west axis centering on London to a north-south one which gradually came to focus on New York City as the economic capital of the United States of America.

If the readily available network of waterways retarded roadbuilding in Virginia, it also made bridges and ferries a necessity. As early as 1641 a law was passed providing for these to be erected by the counties with funds to be raised by the annual county levy. This arrangement tended to spread the costs over the people of each individual county and evoked many protests. As a result ferriage rates were set by the Assembly in an act of 1647 and the burden was thus shifted to those who were actually using the ferries.

Though the practice became common only toward the end of the following century, the first road financed by a public levy of the colonial government occurred in 1691 as a defense measure. This levy was for the construction of a road to connect the chain of frontier forts located along the fall line of the rivers and allow communication between them. The Indians gradually retired beyond the Blue Ridge as a result of this continuing pressure from the white man and his minions.

By the turn of the century the Indian menace was subsiding in the Piedmont and settlement proceeded apace. The first quarter of the eighteenth century would see the establishment, under the impetus of Governor Alexander Spottswood,of the two new frontier counties of Brunswick in the south and Spotsylvania in the north. This move into the Piedmont would bring the problems of land transportation to the forefront in Virginia.

Transportation difficulties increased disproportionately above the fall line. Previously, navigable water had always been available a short distance away. Now, in many cases the nearest navigable water was avail- able only at the fall line. Though many streams were later cleared for navigation, many others would always be unsuitable. Below the Appomattox in Southside Virginia, streams flowed into the Staunton or Roanoke River, which led away from Virginia and into North Carolina. In fact the real development of a road system in Virginia began with, and was a response to, the transportation problems caused by the settlement of the Piedmont and Southside portions of the colony. Likewise, the development of the later turnpike system under the guidance of the Board of Public Works was a response to the more complicated engineering problems and the greater financial resources required to cope with ,the crossings of the Blue Ridge and the Allegheny Mountains in the late eighteenth and early nineteenth centuries.

With this movement of settlement into the uplands, there was a general revision of the road laws in 1705, with supplementary acts periodically up to the Revolutionary War. County courts were to divide the counties into road precincts and annually to appoint surveyors of the highways in these precincts to be responsible for "making, clearing and repairing the highways. ..for the more convenient travelling and carriage by land of tobaccos, merchandise, or other things within this dominion" and to provide roads "to and from the city of Williamsburg, the courthouse of every county, the parish churches...public mills, and ferries. .. and from one county to another". Bridges and causeways were now to be constructed where necessary to further these ends. Since special skills would often be required, the counties could let contracts for the work, the expense to be added to the annual county levy. Where more than one county was involved in the construction of a bridge, commissioners could be appointed by each county to cooperate in letting the contract and the costs would be allotted proportional to the number of titheables in each county. Spotsylvania and Louisa Counties jointly constructed just such a bridge over the North Anna River in 1748. Thus far, the early bridges have not been studied and it is not known whether there is extant sufficient information on which to base such a study, but a later publication in this series will attempt to deal with this subject.

In the eyes of the law, mill dams were themselves considered to be bridges and had to be maintained at a minimum width of twelve feet to serve as roadways. This was no doubt considered a legitimate quid pro quo for the county court's favor in allowing the erection of the dam across the stream. Incidentally, it might be mentioned that this law is still on the books. Presumably a mill owner could be presented to the court for prosecution today just as Colonel William Randolph, the owner of Dover Mill in Goochland County, was presented to the Goochland County Court in 1734 for failure to keep his mill dam in good repair and "ten ,(sic) foot wide at top".

Surveyors of roads were empowered to pay for and use timber growing along the highways for construction such as bridges and direction posts, the latter of which were mandated by a law of 1738 to be erected "where two or more Cross-roads or highways meet" with "Inscriptions thereon in large letters directing to the most noted place to which each of the said Joyning roads leads". Some surveyors evidently used a very minimal definition of the word "post" for when the act requiring the posts was entered in the Albemarle County Order Book on 27 June 1745 o.s., it was directed that the posts be "at least ten feet from the Ground", about the right height to be conveniently read by a man on horseback. From the number of entries for .payment for erection of these, those presentments for failure to do -so, and the frequent mention of "the Sign post" in road orders it would appear that a fairly good system of signs was available to the traveller along main roads by the latter half of the eighteenth century. Most of these direction posts were probably similar to the one at which Dr. Syntax is gazing on the print by Thomas Rowlandson reproduced on the back cover of this publication. Many stone markers dating from the nineteenth century survive, but, thus far, the only earlier one is a magnificent example, dating from 1794, located in Brunswick County on the old stage route from Petersburg to Halifax, North Carolina.

The ferry system also came in for changes early in the century. In 1702 the General Assembly took supervision of these away from the county courts and thereafter locations and rates were the prerogatives of the legislature. Ferries became, of course, a necessity once settlement reached the area where boat ownership was no longer common as it had been in the Tidewater area. With the vast counties of the frontier, people had to travel long distances across mountains and rivers to transact legal business at the county seat, and still greater distances to reach shipping points at the fall line such as Petersburg, Richmond, Newcastle, Hanovertown, Fredericksburg, Dumfries and Alexandria. Detailed instructions eventually were published concerning proper boats and rates for the various categories of ferriage such as tobacco hogsheads, livestock, waggons, wheeled coaches and riding chairs.

As the development of the Piedmont and the Southside continued, settlement west of the Blue Ridge began about 1730. Though at first a somewhat groping movement by a few individuals, as it gained momentum it attracted speculators such as Jost Hite, James Patton, William Beverley, and the various land companies put together by members of the Tidewater gentry intent on further increasing their fortunes. These men brought settlers from Pennsylvania, from East Virginia, and even directly from Ireland. Once these settlements in the Valley got beyond the frontier stage a new dimension in transportation problems in Virginia became apparent. Hemmed in from eastern Virginia as they were, and with strong ties with Maryland and Pennsylvania from whence they came, their principal trade route developed along the line of the "great Waggon Road" up the Valley to Philadelphia, and later Baltimore. This road had been laid out in 1745 by James Patton and John Buchanan as a result of the Treaty of Lancaster, which promised the Iroquois a marked path up the Valley. This "Indian Road" later became the Valley Pike, and, still later, Route 11. Interstate 81 follows much the same route today.

Nevertheless, there were on both sides of the Blue Ridge, partisans of closer trading links with the Tidewater. In Augusta County John Lewis actively promoted such links in the late 1730's, and is said to have surveyed the road over the mountains into Goochland (now Albemarle) County. This was probably the famous Three Notch'd Road from Staunton through Wood's Gap to Richmond, though he also may have been responsible for the later one through Rockfish Gap. By 1748 these advocates of eastern trade were able to secure the first specific, local road legislation of a non- military nature from the Virginia General Assembly. This first enactment was .to allow the court of Prince William County to make a levy on its titheables to clear a road from Pignut Mountain (presently in Fauquier County) to the Blue Ridge at Ashby's Gap. Almost immediately the legislature followed this with an act appropriating £100 to improve the crossings of the Blue Ridge at Swift Run Gap (Route 33 in Greene County) and at Wood's Gap (now usually called Jarman's) where the Three Notch'd Road passed from Albemarle County into the Valley.

Events in the west were now moving rapidly toward the opening of hostilities with the French over the lands and fur trade of the Ohio Valley. With the French and Indian War, military necessity was to give the greatest impetus thus far to transmontane roadbuilding. As early as 1750 a trail called Nemacolin's Path was marked from Wills Creek near the present Cumberland, Maryland, to Redstone on the Monongahela River. A few short years later this served as George Washington's route on his way to Great Meadows in 1754. The following year it also served as General Braddock's route to his disastrous defeat near Fort Duquesne. The staging of his expedition from Alexandria to Winchester did much to help open the roads passing through the Blue Ridge at Ashby's Gap and at Snicker's Gap.

Braddock's defeat opened a Pandora's box for the Virginia frontier. General Indian warfare shortly broke out all along it, although not all the Indians sided with the French in this conflict. To counter the threat from this quarter a chain of forts was constructed at strategic locations along the Allegheny Mountains from North Carolina to the Potomac River. Since all military supplies had to be brought from eastern Virginia, better roads were imperative if these frontier garrisons were to be adequately supported. Besides the development of roads in the Winchester area, necessity also forced the opening of a road from the area of modern Roanoke (then Big Lick) down the Valley to the Long Island on Holston River where Kingsport, Tennessee, is now located.

All these wartime developments only served to stimulate further demands for roads from the people in these areas at the cessation of hostilities in 1763. Now there were petitions for improvement of navigation on the Potomac River and for improvements on the roads leading from Fredericksburg and Alexandria to the Blue Ridge. Farther south, other petitions asked for roads through the Blue Ridge at Rockfish Gap, Swift Run Gap and a number of other places during the years 1761-1765. These petitions generally met with a favorable response from the Legislature and it is difficult to predict how much might have been accomplished had the Revolution not intervened. Bills were passed for roads through the afore- mentioned Rockfish and Swift Run Gaps, for a road to Pittsburgh along the line of Braddock's Road, and for a road from Warm Springs to Jennings Gap, both then in Augusta County. There were also bills for repairs to specific roads due to the inability of the counties to finance them. A spate of bills in 1772. made further provision for many of these same roads.

All of this special legislation, of course, was beyond the province of the general road law and its system of maintenance by labouring male titheables supervised by surveyors of roads. Some projects were financed by special county levies, others were established as toll roads with trustees in charge; one was financed by the colonial government itself from its own funds. The encounter with the mountains was making more and more apparent the fact that the county road system was simply inadequate to the task before it. Unfortunately, these were the years of growing tension with the mother country and the political energy of the gentry increasingly was diverted from internal improvements to questions about the nature of government itself. With the outbreak of war in 1775 matters of road development and internal improvements were thrust into the background for the duration of the war. The Virginia constitution of 1776 made no mention of any integrated plan for statewide internal improvements. Though the constitution as adopted was substantially the work of George Mason, the draft by Thomas Jefferson, which arrived too late to be considered by the Committee drafting the constitution, is also silent on this point. This seems strange in light of Jefferson's remarks on transportation routes in his <u>Notes on the State of Virginia,</u> written a scant five years later in 1781; however, one must remember what had transpired during the intervening period. The "colonials" of 1776 had become "continentals" by late 1781 with the necessity of peopling half a continent and then binding its transmontane area to the eastern seaboard. Indeed, this was seen as one of the major problems of America into the 1820's, and was partially responsible for the attention internal improvements received during those years.

To say that the Virginia constitution of 1776 was silent on the subject is not, however, to say that nothing at all was happening on this front in these years, for by the end of the war Virginia stretched to the Mississippi River and contained the present states of Kentucky, West Virginia, Ohio, Indiana, Illinois, Michigan and Wisconsin. Although most of this area was later ceded to the United States, Kentucky remained, until 1792, and West Virginia, until 1863, a part of Virginia. In 1779 a road, to be paid for by the state, was ordered constructed into the "County of Kentucky" for the use of the settlers on the waters of the Ohio. Another enactment the following year (1780) allowed Greenbrier County to levy funds to open a wagon road leading from Lewisburg to Warm Springs on the wagon road at the mouth of the Cow Pasture River.

What sort of an assessment can be made of the roads of "the Antient Dominion of Virginia" down to the year 1776? To make such a judgment some comparable political unit (i.e., one with similar people, institutions, topography, problems, etc.) is necessary and, fortunately, several are readily available. Although somewhat lacking in a western area and its attendant problems, Maryland is in many respects similar. Its road history is almost a duplicate of Virginia's. Having the same money crop, it also had the same kind of waterways available. As expected, roads were there administered by the county court through road overseers and their gangs of titheables. These roads also tended to develop as transportation links to navigable water though there were large numbers of causeways, bridges and ferries. There, too, road administration was probably inefficient but probably adequate to the needs and desires of most of the people of the time. Above all it was probably the only system possible, considering the means available at that period. It is probable that these same' generalisations could be extended to the other colonies were detailed studies of their roads available.

Virginia's roads have been compared unfavourably with those of the mother country, England, in the light of the swift coaching system devised there during the nineteenth century. However, if one extends this comparison to all of Great Britain and concentrates instead on the seventeenth and eighteenth centuries the contrast is considerably lessened. Certainly in the earlier period many of the English roads must have been atrocious, with the packhorse the favored animal for conveying goods and horseback the usual mode of travel. Until after the General Highway Acts of 1766 and 1773 tolerable carriage roads were probably as scarce there as they were in America. These acts incorporated what was, for the time, a rather novel idea; that of improving the roads rather than merely maintaining them. They also gave some attention to reducing wear on road surfaces by specifying rim types, weights and sizes of wheels for the different types of carriage.

When the comparison is restricted to this earlier period Virginia's roads do not fare badly at all. Particularly is this so for the Tidewater and Piedmont areas; if the Blue Ridge and Allegheny areas were difficult of access, so were such mountainous areas of Britain as Scotland and Wales. In fact, some of the two last named areas became accessible by road only in this century. Indeed, Virginia's roads probably served well enough the needs of the colonial Virginian riding horseback to and from his tobacco plantation. Few owned carriages anyway, and the transportation of tobacco was usually by hogsheads or waggons drawn by slow-moving oxen. Construction and maintenance of roads satisfactory for these purposes was therefore both simple and inexpensive, well within the reach of the county road system in the Virginia Tidewater and Piedmont. The encounter with the mountains, however, presented difficult engineering problems which were beyond the skills and the financial resources available at the county level. The rapid development of the Valley made roads through the mountains a necessity just as it was becoming increasingly clear that the existing road system could not provide them.

NOTES:

GROPING FOR A SOLUTION
1783-1816

GROPING FOR A SOLUTION
1783-1816

With the close of the Revolutionary War in 1783 the former "colonials" awoke to a changed state of affairs. Now become "continentals", they must now learn to think "continentally". No longer able to depend upon Great Britain's trade and her financial resources, they were also deprived of the umbrella of her military and naval power. Possessed of half a continent divided by a spine of nearly impassable mountains into eastern and western areas with indefensible frontiers under a weak government, it appeared that even if the confederation were to persevere it must ultimately split into two republics, one served by the eastern waters and the other by the mighty Mississippi and its tributaries. The welding together of the nation was to consume the years up to and through the War of 1812. The matter of the weak central government was dealt with by the Constitution of 1787 and fought out during the 1790's and the first decade of the nineteenth century. Tying the disparate sections together by means of an adequate transportation network was to take longer, not being fully achieved until the advent of the railroad in the 1830's and 40's. Until then, water and road transportation were to be the focus of this effort to avert the creation of a "western republic".

Although this same thing later occurred in microcosm in Virginia, when West Virginia was created in 1863, it was not because of a lack of effort to improve means of transportation between the two areas during the intervening period. In 1782 the General Assembly passed a bill calling for a general survey of roads running through the Blue Ridge from the various port towns. Although an admirable expression of sentiment on the part of the legislature, the bill made no provision for funding, it evidently being expected that the surveys would be executed and paid for privately. Not surprisingly this act was ignored by the counties. It remained for George Washington to initiate the first significant plan directed toward the comprehensive development of transportation routes to the West.

There appeared to be three principal routes through the mountains to the Ohio-Mississippi Valley: (1) in the north, along the line of the Potomac- Monongahela; (2) through central Virginia, the James-Kanawha water route; and, finally, (3) the land route by the Wilderness Trail through Cumberland Gap near the border with North Carolina and Tennessee. In 1784 George Washington travelled to the trans-Allegheny area looking for the best route to connect the Potomac River with the Ohio River. After his return he communicated to the Governor of Virginia plans for one connection using the Potomac River and for another using the James. He also pointed out the fact that the states of New York and Pennsylvania were making efforts to secure a similar transportation link with the West in hopes of tapping its trade. Washington's action resulted in a resolution from the House of Delegates proposing a "survey of the rivers James and Potomac to their several sources... and how far and for what distance the land carriage may be reduced to, between the said rivers and the western waters". This resolution, which led to the incorporating of the Potomac Company and the James River Company that same session, was the beginning of the internal improvements movement in Virginia. Initially devoted only to canalising their respective rivers, the Potomac Company managed to build canals around the falls near Georgetown and the James River Company constructed one around the falls at Richmond, with little being done by either company to connect their respective rivers with the western waters. Later, in the nineteenth century, these companies were to grow into gigantic canal companies with tributary river navigations and turnpike companies, but as work continued on the canals at the falls it became obvious that the grand design of a connection with the west could be achieved only with a vast expenditure and great effort.

Nevertheless, the hue and cry for better communication continued. Trans-Allegheny Virginia was rapidly being settled and the old county road system was less and less able to meet the demands now being placed on it. Particularly was this so regarding the demand for waggon roads from the West down to the Tidewater ports. The necessity for these was beyond question, and it now began to force the evolution of a systematic plan for statewide internal improvements in Virginia. This evolution was to be very gradual, however, lasting up to and through the War of 1812, during which time a piecemeal approach was tried with state aid for local roads and with turnpike companies being chartered and lotteries resorted to in the hope that private means would suffice. Over the years from 1785-1816 this uncoordinated, shortsighted, piecemeal effort gradually gave way to a vision of a state-controlled system of internal improvements to be formulated, implemented, and coordinated by a state body created specifically for that purpose and funded by the General Assembly.

A revision of the general road law in 1785 was one of the first steps in this piecemeal approach. This, it was hoped, would improve coordination throughout the state. Provisions of previous acts of 1748 and 1762 were tightened up and modernised. Special committees to study road conditions ~ and report on them would be appointed. Writs of mandamus were now to be available to force recalcitrant county courts to engage in joint projects to build structures such as bridges and causeways where necessary. In recognition of the hardship that the mountainous terrain imposed on those trying to maintain roads, an amendment of 1789 to this act permitted seven of these western counties to maintain what were called "expedient" roads only; that is, roads cleared and smoothed to a width of thirty feet. All of this is not to say that any real changes were effected in the old system by these acts. With the average road overseer being in charge of a portion of road probably less than five miles long within a single county, even the maintenance and improvement to some fixed standard of a road across a single county involved the gentlemen justices of the county court, some four or five overseers and an indeterminate number of labouring titheables to do the actual work, so that almost impossible problems of supervision were presented. To take a hypothetical example, the famed Three Notch'd Road from Richmond to Staunton, it can be seen that the counties of Henrico, Goochland, Fluvanna, Louisa, Albemarle, and Augusta would have been involved in any effort to bring this road up to the level of a turnpike or all-weather waggon road. Six county courts (with possibly as many as 20 justices each) and probably 35 or 40 road surveyors and their titheables, as well as a plethora of special bridge commissioners, commissioners for inter-county cooperation, etc., would have been involved for a total of 150-200 people having to do with supervisory or management functions and many, many more actually involved in executing the work. In point of fact, as time went on more road surveyors tended to be appointed with smaller segments of road to maintain so that in the older, eastern counties the number of people involved would have been still greater.

Fortunately, the construction of the turnpikes and through roads was not attempted by this method. First efforts in this direction tended to take the form of special grants of funds or tax arrears to be expended on specific roads or segments of roads. Occasionally, roads which passed through several counties had commissioners appointed by the state to lay out and supervise the improvement of the road. These men reported directly to the General Assembly although the costs were usually borne, and the work executed, by the actual counties concerned. Most of the recipients of this aid were counties in the trans-Allegheny or Southwest Virginia, and the usual purpose of the aid was to complete portions of roads which connected these areas with eastern Virginia. Probably the most significant effort by the state was along the line of the James and Kanawha Rivers. Increasingly, this route across the heart of Virginia was becoming the focus of her program of internal improvements. In 1802 a waggon road was ordered constructed from the head of navigation on the James River to the head of navigation on the Kanawha River in present West Virginia. The General Assembly appointed a road commissioner, who would report directly to the governor, to oversee the surveying, location and construction of this road. Tollgates were established in 1809 in an effort to make the road self-supporting, and in 1813 the road was divided into three segments, with each to be under a separate superintendent.

The toll road was one method of coping with the inadequacy of the county road system. As early as 1772 toll rights had been given to Augusta and Nansemond Counties to help them keep up two specific roads. In 1785 the Assembly appointed commissioners to set up turnpikes on the roads from Alexandria to Vestals' and Snicker's Gaps to keep them in repair. The road from Ashby's Gap to the Snicker's Gap road was made a turnpike in 1787 as was the one from Chester's Gap to Richmond, The heavy waggon traffic con- verging on Alexandria and Richmond made it simply impossible to maintain the roads any other way. The heavily travelled roads through the gaps in the Blue Ridge at Swift Run Gap (1802), Thornton Gap (1806), Rockfish Gap (1808) (See Douglas Young, A Brief History of the Staunton & James River Turnpike, Virginia Highway & Transportation Research Council, 1975), White's Gap (1808), and Milan's Gap (1814), as well as the roads through the mountains of Southwest Virginia into Kentucky, became turnpikes in the ensuing years.

Besides these more or less public turnpike companies, beginning in 1795 a number of private companies were enfranchised by the Assembly. The first of these was for a turnpike from Alexandria to Little River in Fairfax County. Reorganized in 1802, it completed a 20-foot-wide road 34 miles west from Alexandria in 1811. Though the "turnpike era" in Virginia would probably be better dated from 1815 onward, there were eighteen charters granted turnpike companies prior to the War of 1812. Only about half of these survived, and ! all of these were located either in the Richmond area or the heavily travelled Winchester-Alexandria-Fredericksburg area of Northern Virginia.

The specifications for construction in these acts varied considerably. Widths specified ranged from thirty to sixty feet according to the terrain and the anticipated loads to be hauled over the road. After 1810 the state also attempted to establish maximum permissible grades for each road in the enabling act.

Concurrently with the development of the toll road method of financing roads, lotteries were also being used. In at least one case, that of the Staunton & James River Turnpike through Albemarle and Augusta Counties, a road initially constructed by means of a lottery was later improved and be- came a turnpike. The usual lottery arrangement was supervised by the state through commissioners appointed by the Assembly to do the advertising. selling of tickets, drawing, and the distribution of the prizes in return for ten percent of the proceeds. Most lotteries, however, were in the Allegheny or trans-Allegheny areas, where the county road system had never been, and from the nature of the terrain could never be, too strong. Both the toll road and the lottery were methods of bypassing the counties to achieve desirable objectives to which the state was not yet ready to commit itself whole-heartedly.

Progress was agonisingly slow as the need for a body to coordinate long-range planning slowly became evident. Virginians watched while the commerce of the western part of the state was siphoned off by Baltimore and Philadelphia. Local control of roads was a prerogative still jealously guarded by the Virginia gentlemen sitting on the benches in the classic red-brick court houses of eastern Virginia and, indeed, by many of those in the newly settled regions further to the west. The apparent laggardliness of the state led many to begin to look toward the Federal government for aid in developing internal improvements and to continue looking in that direction even after the state did become active itself. Interest in Federal aid to internal improvements was to be one of the sources of conflict between eastern and western Virginia during the ante-bellum years. For the most part though, this conflict occurred later; during the period of groping prior to the War of 1812 the focus of internal improvements sentiment still remained local. Movement away from local control, however, was slow but inexorable. Within limits the new methods of financing roads were working: publicly owned toll roads were followed by private turnpikes and lotteries as a network of tolerably good waggon roads was established to serve Richmond, Alexandria, the Blue Ridge and beyond. Canals now allowed batteau navigation by the falls of the James and Potomac Rivers. Although from the success of these experiments it appeared that the means for joining them were at hand, eastern and western Virginia remained separated by the mountain barrier. As the maturation of the public's thinking on this subject approached, comprehensive schemes began to be broached by the newspapers, always leaders in projecting for progress. In 1804 the Richmond Enquirer, the mouthpiece of the then dominant political organisation called the "Richmond Junto" of whom its editor, Thomas Ritchie, was a leader, suggested the problem could be resolved by the construction of a good waggon road from Richmond over to the falls of the Kanawha River, to be built and maintained by the state itself. This road would have leading in to it a series of feeder roads from the Roanoke Valley, the New River Valley and that of the Holston. North of the James River state roads would be constructed from Staunton, Swift Run Gap, and Winchester by way of Culpeper Court House and Hanover Court House to Richmond. Another road would run from Alexandria by Fredericksburg to Richmond. The keystone of the arch would have been a road connecting the three commercial towns of Richmond, Petersburg and Norfolk. The choice of the James-Kanawha route was indicated by the fact that the James was already navigable down to Richmond by batteau, and the connecting road between the heads of the rivers was already owned by the state. Agitation continued, via newspapers and the stump, for better roads as relations with England worsened and the spectre of war raised its head. Governor John Tyler called for internal improvements in his annual message to the General Assembly in 1810. Henry Banks, a tireless promoter of internal improvements, linked better banking to road improvements in 1811 in a book he published at Richmond called <u>Sketches and Propositions Recommending the Establishment of an Independent System of Banking, Permanent Roads.</u>

Finally, in February 1812, state action came when the Assembly appointed twenty-two commissioners to view the James River from Lynchburg west, the New River, the Greenbrier, and the Kanawha River and report on the practicability and the probable expense involved in making these rivers navigable. Three of the commissioners were directed to find the best route for a turnpike from where Dunlop Creek entered the James River over the divide to the Greenbrier River.

That fall a resolution proposing a "Fund for Internal Improvement" for the purpose of making navigable the principal rivers, particularly the eastern and western waters, and connecting. them by public highways, was presented. This resolution, providing for a Board of Public Works to administer the fund, was tabled because of the War of 1812, but it did lead to the setting up of a House Committee for Roads and Internal Navigation with responsibility for determining the internal improvement policy of the state. Shortly, a coherent transportation program would emerge as a result of the establishment of this committee.

NOTES:

THE BOARD OF PUBLIC WORKS
1816-1827

THE BOARD OF PUBLIC WORKS
1816-1827

Lessons learned at the national level during the War of 1812 were not lost on the Virginians. There, besides problems made evident in financing the war effort, difficulties in moving troops to desired locations in the country demonstrated the need for improved transportation facilities in the national interest. Farsighted individuals at the national level, such as John C. Calhoun, the Secretary of War, and others, would shortly begin a movement for federal internal improvements, only to be stymied by James Madison's recollection that this power had been discussed and decided against during the Constitutional Convention of 1787. Therefore, thought Madison, the Federal Government had no power to commission any internal improvements.

At the state level, however, the question was one of will rather than of constitutional prohibitions. The principal objection of the believers in state sovereignty and state's improvements was to Federal internal improvements rather than internal improvements themselves. Federal internal improvements was an issue viewed as the opening wedge in an assault on the states themselves, as the first of a series of federal encroachments on the powers of the states.

In the afterglow of the War of 1812, during the period of "generous nationalism", many states were active in the field of internal improvements. Probably the most famous was New York's Erie Canal, but the other states, although not possessed of such an excellent route as New York's were not idle. In Virginia, the action taken by the General Assembly in 1812 had been con- firmed by the reactivation of the Committee on Roads and Internal Navigation in 1814, and its issuance in 1815 of a long report dealing with the transportation needs of the state.

This report, which laid heavy emphasis on the use of improved engineering techniques, made Virginia's future dependent upon the development of a transportation network, connecting the eastern and western sections of the state together. Taking a long look at the financial condition of the state, it predicted that increasing revenues would serve to make the project successful. It went on to recommend that a fund be created, to consist of stock already owned by the state in various river navigations, banks, canal companies and turnpike companies, and that the income from these to be used to subsidise the various internal improvement projects. Following the report's recommendations, the Virginia General Assembly, on February 5,1816, passed an act to create this fund and to establish a Board of Public Works to supervise the use of it.

As originally constituted the fund totalled more than a million dollars, with the promise of any future stocks or bonuses which might be received by the state for the incorporation of new institutions or increasing the capital of existing ones. To supervise the expenditure of the income from these stocks a Board of Public Works consisting of thirteen men was appointed. This board, the forerunner of the present Virginia Department of Highways & Transportation, was to be headed by the Governor and consist of the following: the Treasurer and the Attorney-General, three members from west of the Allegheny Mountains, two from between the Allegheny Mountains and the Blue Ridge, three from between the Blue Ridge and the fall line of the rivers, and two from the area between the fall line and the coast. Thus, the interests of the varying sections would all be represented on the policy-making board. The board was instructed to invest the income from the fund in selected companies engaged in public works. Investment was allowed in a company by the state after a minimum of three-fifths of its stock had been subscribed to by private individuals. A principal engineer who would make surveys and super- vise designated projects was to be appointed by the board.

Another product of this burst of legislation was a scheme, later revised, to accurately map each county and, using these maps, to construct a large map of Virginia. In the 1820's the immense nine sheet map measuring eight by thirteen feet would be published by the state of Virginia. Corrected and up- dated this map was reissued by the state, and in a reduced size is available from the Virginia state library as a part of E.M. Sanchez-Saavedras' masterful work A Description of the Country 1607-1881, a great boon to those interested in Virginia history.

Capping off a very productive session the legislature chartered ten new turnpike companies, nearly twice the number of any previous year. Reaction to the accomplishments of the General Assembly was favourable, with Thomas Ritchie's Richmond Enquirer praising the Board of Public Works' plan in editorials on March 2, 9, and 23, 1816, as one combining vigilance, wisdom, and capital successfully.

Meanwhile, at the national level, Virginians were not inactive, as an examination of the Congressional Record will attest. Virginia's state aid being dependent on the amount of private capital subscribed to private projects in a given area, the undeveloped trans-Allegheny tended to be short- changed because of its lack of available private capital. As a result, congressional representatives from this area tended to favour federal aid to internal improvements. Ballard Smith, of the Kanawha district, proposed that the Federal Government be allowed to invest in whatever company Virginia set up to complete the James River and Kanawha system. A number of petitions received from his constituents had impelled him to this action. While others were similarly motivated, this trend was soon headed off by action taken by the Virginia government itself.

An act passed by the Assembly in February 1817 prescribed the regulations for the incorporation of turnpike companies. The Assembly would now grant charters to companies to construct roads, would fix their capital stock, name their commissioners and designate the places in which the stock would be offered for sale to the public. Capital stock was to be divided into shares valued at no more than $100 each by the commissioners, and when half this stock had been subscribed they were to authorize the subscribers to elect officers consisting of a president and five directors. These, once elected, were to have power to appoint the administrative officers, let contracts for construction of the roads, and, of course, convene meetings of the stock- holders as well as to receive further subscriptions. As soon as an amount to the value of three-fifths of the capital stock of the company had been subscribed by private parties, the

Assembly could have the Board of Public Works subscribe to the company with the Board receiving a proportionate voice in the management of the company.

This act also set out regulations for the construction of the roads, specifying that all water courses crossing the road should be bridged where necessary, that the road should be sixty feet wide at least, that eighteen feet of this should be covered with gravel and stone and kept free from mudholes and ruts so as to be usable for carriages and heavily laden waggons, and that on either side of this part of the road a "summer road eighteen feet wide should be kept in good repair for the use of waggons and other carriages in dry weather, between the first day of May and the 31st day of October. . ." Wheel sizes were specified, as well as maximum weights of loads. Toll gates were to be erected, but if the road were out of repair, tolls ceased until such time roads were completed. Toll rates were also specified in the act. Turnpike companies were to be allowed two years to begin construction and ten years to complete it. Failure in either of these particulars was to result in the forfeiture of the company's charter. Each company was directed to file a report with the: Board of Public Works at the end of the first year, and every three years thereafter, for the purpose of toll regulation.

In November 1816 Loammi Baldwin was appointed Principal Engineer for the state. In 1818 he resigned and was followed in the position by Thomas Moore; Moore lived only a short time, however, dying in 1822. Finally, there came to this position the man who would leave an enduring mark on Virginia's internal improvements. This was Claude Crozet, a French engineer then serving as professor of engineering at West Point. In Crozet the Board of Public Works finally got the man for whom they had been searching, a brilliant engineer able to direct Virginia's long-range program of internal improvements. Crozet's appointment also nipped in the bud a movement which was then afoot to abolish the Board itself.

Over the next three decades Virginia's internal improvements system was to grow and prosper under the guiding hand of Crozet, though not without occasional conflict within the General Assembly between the proponents of canals, turnpikes, and railroads. Knowledgeable in the methods developed by John McAdam-and Thomas Telford in England, Crozet urged their use on Virginian roads. It took awhile but he was finally able to show the turnpike companies and the counties than the paved road was in the long run the lowest cost highway which could be built.

It might be well here to briefly delineate the ideas of those two Scottish engineers, Telford and McAdam. McAdam's definition of a road was that it was an artificial floor, and that it should therefore be strong enough for the passage of any wheeled vehicle regardless of weight. A well- drained, dry roadbed was of primary importance to his conception of a road. To ensure this, his roadbeds were elevated above ground level, with paved drains issuing from the lower level. On the foundation stones of the bed was laid a surface of clean, dry flint stones. McAdam recommended that both the surface layer and the roadbed itself be composed of broken stones no more than an inch in diameter. From experimentation he had discovered that stones larger than this tended to work upward under the constant pressure and vibration induced by heavy traffic.

Thomas Telford favored a somewhat different approach to road building, wherein the bed was composed of large stones placed by hand to form a close-knit pavement. Upon this base a layer of smaller stones of 1 1/2 -2 1/2 inch diameter was spread and then compressed by a steam roller. The road, when complete, was about eighteen inches thick at the center, tapering to nine inches on either side.

Laommi Baldwin, the first principal engineer, attempted to publicise the idea of the paved road in 1817 in his annual report to the Board of Public Works. His specifications therein called for:

> A road bed thirty feet wide into which are placed large stones well beaten close to each other over the whole width. Upon this is another bed of stones broken to the size of about four inches, well hammered and rammed in, so as to fill all the cavities between the under stratum of large stones. The third and last layer should be coarse gravel or stone broken to the size of hickory nuts, thrown on evenly or rammed or rolled with a heavy iron roller. The first should be from a foot to eighteen inches thick, the second 12 inches, and the last about 10 inches in the middle and 8 at the sides.

These specifications never became law, and it is extremely unlikely that any of the turnpike companies would or, for that matter, could have complied with them. Most companies had great difficulty meeting even the requirements of the turnpike act, and had to be granted exceptions in the charter or else in a subsequent act amending the charter. The Acts of Assembly are replete with such acts allowing road widths of a mere fifteen or twenty feet and dispensing entirely with the "summer roads" on either side. Particularly was this the case in mountainous areas where construction was so very expensive.

Nearly a score of turnpike companies were chartered by the Assembly in its 1816-1817 session as a result of the new policies of the state. Ritchie's Enquirer, mouthpiece of the ruling "Richmond Junto", indulged in another piece of self-congratulation on February 28, 1818, calling the act "one which guards the rights of the public by well-digested public regulation". In truth, however, somewhat the same feeling was exhibited by investors in the new turn-pike companies, some of whose projects were quite ambitious. Probably the most ambitious proposals advanced were those of the Northwestern for a road from Staunton to the Ohio River, the Southwestern for a road from Lynchburg to the Tennessee line, the Valley for a pike from Salem to Winchester, and the Winchester & Northampton, which projected a connection between Winchester and a point on the Cumberland Road near the foot of the Allegheny Mountains. Unfortunately, these projects proved beyond the capabilities of private enterprise at the time. The Panic of 1819 and the ensuing depression provided an unfavourable climate, and all of these companies failed; in fact, a majority of the early companies chartered by the Board of Public Works failed. Though there was an initial flurry of charters in the years 1817-1819, after the Panic few sought to invest in turnpikes .and the number of charters slowed to a mere trickle in the 1820's, not increasing substantially until the next decade.

The effect of the Panic can probably also be seen in the investment policy of the Board of Public Works after 1819. An amendment to the Board of Public Works Act in that year directed the Board to invest only in such companies as already had one-fifth of their stock sold to solvent persons and to stop payment of state funds to the company should any evidence appear that company funds were not being properly handled.

As a result of this amendment and, no doubt, the caution quite naturally engendered by the Panic, the investments of the Board tended to expand rather slowly during the years following so that only those companies with the best prospects received aid from the internal improvement fund. Projects tended to be located in either of the two market areas of the state, Richmond on the James River or the Winchester to Alexandria and Potomac River region, though there were some exceptions. The Falls Bridge Company was aided in 1820, the Snicker's Gap Company in 1822; in 1823 the Fauquier and Alexandria, the Fairfax, the Wellsburgh (now West Virginia) and Washington, and the Lynchburg and Salem Companies; in 1824 the Manchester and Petersburg Company; in 1826 the Tye River and Blue Ridge (Nelson County), the Falls Bridge again, the Shepherdstown and Smithfield, and the Staunton & James River Companies (Augusta and Albemarle Counties); and in 1827 the North-Western Road Company.

When it came to engineering assistance the same two areas also were favoured, although the James River area was most favoured by the Board of Public Works. In fact, it would probably be better to say the James-River- Kanawha route was most favoured, though the terminus was still the Richmond- James River area. Loammi Baldwin, the first principal engineer, proceeded in 1817 to locate the Kanawha road from Dunlap's Creek (now Covington) to the falls of the Kanawha River, estimating the cost to be $372,000. Further surveys and plans for development along this route appear in the report from the Board of Public Works in 1819. By 1820 it was evident that the James River Company was unable -to execute the grand design envisioned by the state; there- fore, the General Assembly in that year authorised the state to purchase the company from the stockholders. Simultaneously, it was decided to place the Kanawha road under the supervision of two commissioners resident west of the Blue Ridge.

The state had hoped that the old James River Company would open navigation of the river to the mouth of Dunlap's Creek; from there to the falls of the Kanawha, travel would be by the road, and beyond that the Kanawha would be opened for river navigation to the Ohio River, which would funnel the trade of America's hinterland (or at least a goodly portion of it) through the ports of Richmond and Norfolk. Indeed, from 1816 into the 1850's, and especially after the state acquired the Company, this project, as the principal east-west connection, the "keystone to the whole system of internal improvements", absorbed the principal part of the energy and resources of the Board. During the depression subsequent to 1819 the Board advanced the James River Company funds amounting to $800,000 to construct the Kanawha Road, which had been begun in 1822. With these sums the road was completed, a distance of ninety- three miles, and opened for use.

If the James River improvements received- the lion's share of appropriation, other areas still did not lack for attention in the years following Crozet's appointment as Principal Engineer. In 1823 Crozet surveyed Slate River, Patterson's Creek, and the south branch of the Potomac River, filing reports on the prospects of their being made navigable by clearing, installing locks, and making other improvements. He also, this same year, initiated surveys of the roads between Staunton and Parkersburg and Winchester and Romney with an eye to their improvement. In the year following (1824), he examined and filed reports on the Alexandria and Fauquier turnpike, the Monongahela River and the James River through the Blue Ridge. After this he concentrated for three years on the region of the trans-Allegheny, surveying roads from Romney via Clarksburg to the Ohio River, from the Ohio River at Fishing Creek to Morgantown and to Smithfield, in Pennsylvania, besides examining the Kanawha turnpike and making recommendations towards improving it. One of these recommendations called for a road to connect the turnpike with Staunton. A survey and report was also made on a road between Danville and Wythe Court House and one between Lynchburg and Lexington. Several more extensive river surveys were completed, among them a complete reexamination of the James River from Maiden's Adventure in Goochland County all the way to Dunlap's Creek. In line with the requirements for financial stringency, a number of turnpike companies with state investment were critically examined and recommendations for improvements made by Crozet.

While the effort by the state focussed on the turnpike and canal system, the county roads were not wholly neglected. With the codification of the law in 1819 the general road law was revised to meet the conditions that had evolved since the previous revision in the year 1785. Surveyors of roads were given more power to prosecute those labouring male titheables who failed to appear and perform their road work, and the gentlemen justices of the county court were authorised to establish toll roads where they thought them necessary. Modern street-racers might note that the then popular custom of racing horses over a quarter-mile course (or longer distances) on the public highways was outlawed. Besides the attention devoted to the financial aids given to the counties by the state in the form of toll grants, lottery grants and direct appropriations, professional advice from the state's engineers was now made available to the counties on significant roads.

By 1827, after five years in the field of Virginia, Crozet had a comprehensive understanding of the internal improvements problems of Virginia, and in his annual report that year put forward his view of the most needed projects in the state. Recognising the James-Kanawha route as the linchpin of Virginia's internal improvements system, Crozet recommended construction of the following facilities: (1) a road from the Balcony Falls canal, where the James River passes through the Blue Ridge, to Covington (this in lieu of the James River & Kanawha Canal, not yet completed that far); (2) an extension of the Kanawha turnpike on to the Guyandotte and Big Sandy Rivers; (3) a road from Charlottesville to Parkersburg via Staunton; (4)

a water connection between the James and Roanoke Rivers utilising Buffalo River, near Farmville, and Roanoke Creek in Charlotte County; (5) a connection, by either railway or canal, of the Roanoke and New Rivers; and (6) the extension of the Little River turnpike from Alexandria all the way through Winchester and Clarksburg to Columbus in Ohio. Thus, after a decade of its existence, the Board of Public Works finally had at hand a farsighted plan for the connection of the eastern and western parts of Virginia.

Completion of the canal-river navigations and their attendant feeder turn-pikes, along with improvement of the existing county road systems, might move Virginia into a position competitive with New York. Pennsylvania and Maryland for the trade of the area drained by the Ohio River.

Nor was this merely the idle boasting of a group of visionaries and projectors amusing themselves. Crozet's plan, instead, represents the tremendous progress made by the Board during its first decade. Financial aid was now available for responsible transportation corporations, whether turn- pike, river navigation, or canal company, as well as engineering assistance from the office of the state's Principal Engineer. Important long-range surveys had been conducted along the important canal and road routes. Turnpike companies had been established on many of these routes, and these had been assisted by the Board. Many, if not most, of the companies had managed to somehow survive the Panic of 1819 and the ensuing depression. Much had been accomplished along the route of the James-Kanawha toward the Ohio River, and, most important of all, there was now an organisation possessing skilled leadership engaged in both the work and in developing and training more people skilled in road and canal construction techniques.

Nevertheless, Virginia, comparatively speaking, continued to lag behind such progressive states as New York, which had completed her Erie Canal in 1825 and was now busy With the construction of subsidiary feeder canals, besides enjoying the flow of western trade from the Great Lakes area which the Erie brought to New York City. Virginia, of course, possessed no such good route to the west as that of New York's Erie Canal but this did not prevent Virginians from developing an impatient, "immediately-if-not-sooner" attitude toward the state's internal improvement plans, apparently expecting overnight completion of the main east-west link in the system. Another apparently unrealistic expectation of many Virginians was that these projects would enjoy immediate financial success, without allowing time for travel and trade to develop along the particular route. Often, no doubt, private gain was the principal motive behind construction of a bridge or a turnpike,' rather than any desire to serve the public at large. Some of these attitudes, combined with the bitter sectional rivalries within the state, had almost caused the scrapping of the whole program of public works in the early 1820's. With this scrapping narrowly averted, Crozet's plan now seemed a likely antidote to the poisonous vapours of distrust and sectionalism still hovering in the air in Virginia.

With Crozet's program postulated, there appeared on the horizon what might be called the most revolutionary invention of the nineteenth century, if, indeed, any device deserves that appellation. This revolutionary invention was the steam locomotive and the steam railway, which took the form, in this case, of the Baltimore and Ohio Railroad Company. This company, appearing on the scene in Virginia in 1827, ushered in a new era, reviving sectional jealousies and causing hostility on the part of the canal, turnpike, and navigation companies toward the railroads. Ultimately, of course, the steam railway was to spell the doom of the canals, river navigations, and turnpikes, but this was yet only dimly seen by their advocates.

In spite of, or, possibly because of, the steam railway, the next several decades would be Virginia's most expansive period of internal improvements prior to the twentieth century.

NOTES:

THE BOARD OF PUBLIC WORKS:
THE GOLDEN YEARS
1827 -1840

THE BOARD OF PUBLIC WORKS
THE GOLDEN YEARS
1827 -1840

"An act to confirm a law passed at the present session of the General Assembly of Maryland entitled 'An act to incorporate the Baltimore and Ohio Rail Road'" went on the statute books of Virginia March 8, 1827, inaugurating the "age of steam" in the Old Dominion. With the "age of steam" there came new methods, and with them new problems, to the Board of Public Works. The ensuing railroad development was to be checked momentarily by the Panic of 1837 and the hard times which followed, but recovering its momentum, it then continued to 1861.

Progress was not confined to railroads; canal and turnpike construction also reached new heights during this decade. Overnight and nationwide, there now seemingly developed a consciousness of the need for better communications. Edward Graham Roberts cites four significant factors in the development of this consciousness: (1) the great physical distances involved in western settlement, (2) the steam locomotive and the railroad, (3) the astonishing success of the Erie Canal, (4) the fight, at the national level, over the American System of John Quincy Adams and Henry Clay. All these things increased the public awareness of the importance of improved means of transportation, and, public awareness becoming public opinion, groups came into being devoted to lobbying for improved transportation facilities. Although there was agreement as to the necessity for action, there was still much controversy as to the form it should take. In Virginia, internal improvement conventions were held and petitions were submitted to the General Assembly; and newspapers were filled with editorials and letters demanding action as railroad, canal, and turnpike advocates fought it out. Sectional interests added to the controversy occasioned by these differences over technology, and this controversy was further heightened by the growing desire of certain commercial towns to become the principal market for the state. Despite all this, there was. sufficient consensus to allow most of the internal improvements companies in Virginia to survive the Panic of 1837 and its attendant depression.

In Crozet's master plan Virginia had a program which could have readily been adapted to the use of rails, canals, or turnpikes, whichever proved most expedient in a particular case. Unfortunately the program became embroiled in the political controversies of the Jacksonian period, to its great detriment. The matter of the Baltimore and Ohio Railroad revived sectional animosities within Virginia since it raised a hope on the part of western Virginians that the line would be routed via the Shenandoah and Kanawha Valleys to the Ohio, a hope they were to see dashed on the rocks by the opposition raised by the James-Kanawha interests. Simultaneously, the threat of a national system of internal improvements exacerbated sectional rivalries when the completion of the Chesapeake and Ohio Canal became a national project under the General Survey Act. Within Virginia there was generally a division between east and west with the Potomac area and the trans-Allegheny tending to support Federal internal improvements. The fact that Federal internal improvements were usually, though not always, associated with the Adams-Clay American System served to further complicate the picture as the coalition which would depose Adams in 1828 began to take shape at the state and national levels. The states' rights opponents of Federal internal improvements, in control of the General Assembly, proceeded to pass a stinging resolution which denounced the General Survey Act as an unconstitutional infringement upon the powers of the states. One of the leading exponents of the particularist position in this resurgence of states' rights feeling, Governor William Branch Giles, after writing a series of biting articles in the Richmond Enquirer, asserted in his 1827 address to the Assembly that the Federal Government had no power whatsoever to participate in internal improvements.

In fact, internal improvements played a considerable part in the election of 1828. President John Quincy Adams, under the provisions of the General Survey Act, had despatched groups of engineers to just those areas in Virginia where their presence would prove most beneficial to the administration in winning support for Clay's American System. This became a political issue when the new appropriations bill came up for passage. Senator William Cabell Rives, of Albemarle County, led the forces opposed to continuation of this Federal program, attacking Adam's actions as an effort to seduce certain districts from their allegiance to Andrew Jackson in the coming campaign. The Rives group was opposed by the representatives of the transmontane section led by Charles Fenton Mercer of the Loudoun district. This conflict served to further emphasise the increasingly divisive nature of sectionalism in Virginia politics. The east-west nature of the conflict was clearly visible in the election of 1828 as the Rives faction, essentially eastern, supported Jackson and the Mercer faction, more or less representing the west, supported John Quincy Adams.

Even after Andrew Jackson's election, however, the west still treasured the hope that Jackson might continue Adams's policies. Particularly was this the case when, in the Virginia Constitutional Convention of 1829-30, it became evident that not much action was to be expected from the east regarding roads and canals. In spite of the bitter protests of members from the Kanawha and Monongalia districts the Convention virtually ignored the issue of internal improvements. At the national level a bill for a turnpike to run from Buffalo to Washington and on to New Orleans brought Valley interests into the national internal improvements lobby, but eastern Virginians in Congress steadfastly opposed it and after a second reading the bill was tabled. Before the final vote could be taken Jackson's veto of the Maysville Road bill delivered the coup de grace to hopes for federal aid to internal improvements. Once more the issue was returned to the state of Virginia for action.

Though the political and sectional conflicts of the Jacksonian period were preeminent, there were within Virginia many farsighted individuals able to realistically look beyond them toward the future. These men realised that Virginia's economic destiny demanded the fulfillment of Crozet's plans. While the campaign of 1828 was yet in full swing some of these people met in Richmond on May 20, and issued an invitation to delegates from all parts of the state to meet in a convention at Charlottesville on July 14 to consider the internal improvements policy of the state. Bitterly condemning the James River Company for its dilatory policies the convention, in a strongly-worded resolution written by James Madison, petitioned the General Assembly for immediate completion of the James-Kanawha system and its feeders. Other petitions flowed into the Assembly that same year, along with a resolution from a Kanawha County committee, as public pressure continued to mount. Lynchburg joined the hue and cry late in 1830 and in 1831 the sentiments of the western counties were expressed in a memorial from the Lewisburg Convention also demanding immediate completion of the James-Kanawha improvements.

Actually the argument was over the relative speed with which certain segments would be completed, and, while the controversy raged, work continued on all the projects approved by the General Assembly for the Board of Public Works. Petitions requesting the incorporation of turnpikes and other internal improvement companies continued to be received by the Assembly, as well as other petitions praying for state appropriations for these newly formed companies. In the years between 1828 and 1831 more than thirty new turnpike companies were granted charters while the state's engineers were busily engaged in laying out new highway routes. For the year 1828 alone, Crozet .reported surveys for the road from Middleburg to Berry's Ferry and Strasburg, and for the road from Harrisonburg by way of Franklin to Beverley, as well as actively supervising the Staunton and Riffle's Run Road. In 1829 he laid out the Kanawha Turnpike extension from Charleston to Guyandotte, and surveyed another extension from Charleston to Point Pleasant. The year 1830 brought a survey for a road from Covington to Richmond through Lexington, and one for a pike from Staunton and Harrisonburg over to the Warm Springs Mountain Turnpike. Another was made that same year for a road from Lynchburg to Danville. All of the aforementioned surveys, of course, were for roads designed as tributaries of the James River and Kanawha navigation scheme, so that it can be seen that progress was being made. The agitation for more speed seems to have originated in the idea evidently held by some people that this vast project could be completed and all its ends effected "overnight", rather than as the result of years of careful planning and very deliberate work.

As the third decade of the nineteenth century opened in the Commonwealth and the controversy over the speed, or lack of speed, with which the James River-Kanawha improvements were being completed deepened, the whole atmosphere surrounding internal improvements thinking was being altered by the advent of the steam railroad. Previously, canals had been looked to for salvation with the idea that America would repeat the earlier experiences of England with them. There, beginning with the Bridgewater Canal, there had been a steadily increasing employment of this means of transportation throughout the eighteenth century and the first quarter of the nineteenth. Indeed, with the success of New York's Erie Canal it did appear at first that American development would pattern itself after England's.

Now, with the cheers and encomiums still echoing, there appeared the steam railroad. This appeared particularly appealing to the Tuckahoes of eastern Virginia who watched the progress of the Baltimore & Ohio Railroad with the growing realisation that railroads were the wave of the future. Shortly, railroad companies began to be chartered in Virginia itself. Several were chartered in 1830 and 1831; the

Winchester and Potomac, the Staunton and Potomac, the Lynchburg and New River and the Petersburg and the Loudoun. More significant of the changed climate was the fact that the Assembly, motivated by supporters of the James-Kanawha route itself, ordered the Principal Engineer to prepare a study of the practicability of using a railroad in place of the canal and river navigations along that line and in other projects in the state.

The recommendations ultimately made to the General Assembly by the Committee on Roads and Internal Navigation, while giving way so far as to call for a rail link between the James and Kanawha Rivers and the completion of the Petersburg Railroad to the Roanoke River, discarded Crozet's idea for a railroad from near Richmond to Charleston (now West Virginia) and maintained the position that a continuous canal should be built over the whole distance. Other recommendations of the Committee called for a number of state roads to be built. One of the proposed roads was to run through the Valley from Harper's Ferry to the Tennessee line, one from Winchester by Romney and Clarksburg to the Ohio River, one from Staunton and Harrisonburg to the Ohio River, one from Harrisonburg via Fredericksburg to Richmond, one to connect Morgantown with the Cumberland Road, another from Logan Court House to, the forks of the Big Sandy River, and, finally, a road from Lewisburg to Louisa Court House in Kentucky.

The road between New River and the Tennessee line and the Northwestern Turnpike seem to have been priority items, however, and received immediate attention, both being surveyed during the summer of 1831. The Northwestern Turnpike Company, having accomplished virtually nothing since its incorporation in 1824 was now reorganised under state control with $125,000 being advanced to aid in its construction. Canal advocates were, nevertheless, able to delay action on these projects, and to have ordered a resurvey by Crozet along their route as well as a separate survey by Benjamin Wright, one of the engineers on New York's famed Erie Canal. Crozet recommended that a canal .be used from Richmond to Maiden's Adventure in Goochland County and a railroad from there to Charleston in present West Virginia, but this idea was scrapped in favour of Wright's recommendation of a continuous canal to the Blue Ridge.

The principal result of this conflict between canal advocates and railroad advocates was a bill from the Assembly ostensibly designed to reorganise the Board of Public Works but which really served to abolish Crozet's office of Principal Engineer. This action finally led to Crozet's resignation in October 1831. Under the bill the existing thirteen-man Board of Public Works was abolished and a new board was constituted, made up of the governor, lieutenant governor, and state treasurer. Thereafter, engineers were to be appointed for individual surveys and construction projects each year, and the office of principal engineer was abolished. Aid from the Board was also made more difficult to obtain by this act. In fact, the conflict engendered by these conflicting sectional and political interests was actually producing a contracting financial policy as shown by the Assembly's rejection in the

1830-31 session of a bill appropriating two million dollars to aid the newly incorporated internal improvement companies.

These were the years when Virginia should have been forging ahead in the field of internal improvements, the years when the James River and Kanawha route should have been completed and its supplementing by railroads along the tops of the watersheds begun. Instead, a vacillating, desultory policy was pursued, one based on the wishes of the group momentarily victorious in the General Assembly at Richmond. To a great extent the thrust of the internal improvements program as well as the location of the specific projects to be undertaken were determined, during these years, by the party placed in power at the most recent election. Further complicating the differences of opinion held over internal improvements were the battles between the Whigs and Democrats during these years. The coming to power of Andrew Jackson in 1828 revived the party spirit which had lain so quiescent during the two administrations of James Monroe. This revival was amply reflected at the level of state politics in Virginia.

In 1832 and 1833 the advocates of the James River development, led by Joseph Carrington Cabell of Nelson County, dominated the Assembly, forcing through a bill to reorganise the James River Company as a joint stock Company. Not too surprisingly, Cabell became president of the reorganised company and Benjamin Wright, of Erie Canal fame, its chief engineer.

In 1834 and 1835, with the .Whigs victorious, attention still focussed on the James Kanawha improvements, but cries for projects in Democratic areas tended to get short shrift. Following this, the Democrats returned to power for several years and did similarly unto Whig areas, although aid was beginning to be extended to the budding railway companies not associated with the James River development. Some two million dollars went toward construction of rail- roads designed to connect the eastern towns, and the Board invested $300,000 in the Lynchburg and Tennessee Railroad Company. The Baltimore and Ohio Railroad Company, repeatedly denied a route through central Virginia (Whig country), finally received aid when it agreed to a route to the Ohio River which ran through the northwest (a Democratic area). Partisan political considerations continued to playa large part in determining who got which projects and the amount of state aid to be expended on each project.

Despite the vicissitudes of the state's internal politics and the difficulty of getting state aid, petitions for new railroad charters continued to increase in Virginia. Between 1832 and 1839 thirty-five railroad companies were chartered, with the peak of fourteen occurring in the boom year 1836. Only in 1838, directly after the Panic of 1837, were there none chartered. With the retrenchment then necessitated by the lean years following the Panic the railroad companies suffered, but most received state aid sufficient to allow them to survive the depression and avoid bankruptcy. By the mid 1840's they were again flourishing, indeed, they were undergoing a wave of expansion.

Although the conflict of the Jacksonian period surged around the question of whether the canal or the railroad should be the favoured means of transportation in Virginia, and although these same legislators were the ones who ultimately determined the shape of internal improvements policy, the fact was that more money was actually expended on turnpikes than on canals and railroads. This was because turnpike construction was faster, technologically much simpler, and much cheaper than either canal or railroad construction. It also allowed investors to recoup their original investment much earlier. The railroad was new and untried in the judgment of the innately conservative Virginians, many of whom continually expressed doubts concerning it.

It was no wonder then that between 1827 and 1840 one hundred and twenty-one joint stock turnpike companies were chartered in Virginia. Besides these, a dozen other state-aided roads began construction during this period. Indeed, most of the effort of the Board of Public Works was directed toward the surveying, locating, and construction of these roads, some eighty-nine of which lay west of the Blue Ridge. New methods made state participation more readily available to these enterprises. Counties as well as turnpike companies could now avail themselves of direct appropriations, ready loans, and investment subsidies so long as the state shared in the administration of each individual road. Also, due to the problems involved in super-vision, both county and company lottery grants were discontinued by the state after 1831. By the new system the state was empowered to match county funds in a proportion favourable to the county. In 1834 this system was first used in Bath, Alleghany, Pendleton and Pocahontas Counties. The county's portion was payable either in money or labour, with the funds being expended under the supervision of the gentlemen justices of the county court. Where neither profit oriented turnpike companies nor local authorities seemed capable of or willing to construct roads deemed expedient by the Board, it could subsidize them directly with administration being accomplished by a board of directors placed in charge of what were actually state controlled non-toll roads.

From 1832 to 1840 most of the road surveys done were for these roads which were to receive state aid. With the Whigs in power efforts were centered on the New-Greenbrier-Kanawha River valleys since this would be most beneficial to the James River and Kanawha improvement scheme. When the Democrats returned to power attention rapidly shifted to areas under Democratic control. In 1835 and 1836 most of the surveys were in the northwestern part of the state and related to the Northwestern Turnpike and the Staunton and Parkersburg Turnpike, whereas from 1837 to 1840 they shifted to a more nearly equal division between northwestern and southwestern Virginia, although there were occasional surveys elsewhere. As there were now to be more state roads requiring more of the time of the state's engineering staff in supervision of construction and maintenance, that these surveys were made at all is quite surprising. Additionally, an increasing number of railroad surveys were being made, which further absorbed the time and attention of the engineers.

If the immediate focus of the effort altered from time to time the larger aim of the internal improvements advocates remained the same: to connect eastern and western Virginia irrevocably together by means of better transportation. As work proceeded on the improvements along the James-Kanawha route, three other routes were being developed with considerable despatch: the Northwestern Turnpike from Winchester to Parkersburg, the road from Staunton to Parkersburg, and the road from Cumberland Gap to Price's Turnpike, which had a good connection with the Kanawha turnpike at Covington. The location of the Northwestern Turnpike was completed in 1834 and five years later it opened as a stage road. Progress was slower on the Staunton and Parkersburg Turnpike until 1839, when the state assumed from the county authority for the road. The Cumberland Gap Road, remaining under county control, progressed even more slowly. Located in 1832 and relocated in 1836, the road was still far from complete in 1840.

While the construction of the major arteries was the most noticeable development during the period up to 1840, almost imperceptibly an improvement in roads was also occurring at the county level. The condition of county roads, a subject of continued agitation amongst Virginians, now produced another revision of the basic road law. In 1835, a new law overhauled the old system, producing two major innovations: a commission with responsibility for county roads, and the direct county road levy. The road commission was to consist of from two to five members appointed by the county and charged with the responsibility for coordinating and directing all the county's road activities. No longer would the gentlemen justices of the county court have to appoint individuals to "view the way" (survey the route) and supervise construction on each job. Also, by this system the overseers of highways, or surveyors of roads, as they were variously called, received more and better supervision. Previously an indictment for failure to perform their duties appears to have been the only form of supervision exercised by the county court. Usually by the time this had occurred the road in question was in impassable condition. Now, it seemed, there would be continuing supervision of the individual surveyors of roads.

The direct levy allowed the use of a force of hired labourers to maintain roads, instead of relying upon "labouring male titheables" as under the old practice. The passage of this act to some extent reflected the increasing economic sophistication of Virginia with its attendant specialisation of skills. In the seventeenth and eighteenth centuries, when most Virginians were tobacco planters requiring minimal roads and usually having free time outside the growing season to maintain them, the old system had worked tolerably well. Now, with the westward growth of the state and its economic diversification it was becoming inconvenient for storekeepers, blacksmiths, and waggoners to leave their places of business to work on the roads. Indeed, as early as 1785 no less a person than George Washington, in a letter to Governor Patrick Henry, had suggested that the roads be removed from the supervision of the county courts and maintained by contractors.

Unfortunately, there was considerable backsliding on the part of the Legislature here, for, as a result of complaints from the western counties, amendments in 1837 and 1839 allowed counties the option of using either the road law of 1819 or that of 1835. During the next ten years, however, the 1835 commission system became mandatory for all but a few counties, excepted by law, and the road commissioners were required to make annual reports to the county courts. Where deemed expedient special grants, such as extraordinary taxing power, were also made to certain counties to improve their roads.

The 1840's would have seen great progress in Virginia's internal improvements had not the Panic of 1837, with its ensuing depression, intervened. This depression served to throw the new railroad lines back upon a dependence on state largesse, while the proposed line linking Tennessee to the James River was abandoned. Besides' the effects of the depression, the canal versus railroad conflict slowed progress on the James-Kanawha route, and sectional divisions hampered both turnpike and railroad construction. There was also a predictable amount of waste of effort and money due to impractical projects, sloth, and plain mismanagement. After saying all this, however, one still must concede that the 1830's were a period of great accomplishment. East and West were now connected, if imperfectly, by a canal and turnpike along the James-Kanawha line and by a turnpike from Alexandria to Parkersburg. The Louisa Railroad was moving from Richmond up the ridge towards Gordonsville and lines already connected Fredericksburg and North Carolina. Up to the beginning of the policy of retrenchment and consolidation by the state in 1838, Virginia's internal improvements system showed every prospect of success in a few short years. In the face of the crisis caused by the depression, the General Assembly acted to prevent the collapse of what had thus far been accomplished with an 1838 loan act and with another the following year authorising the Board of Public Works to make loans to internal improvement companies and to borrow for the internal improvement fund itself. Efforts to borrow not meeting with much success, the state was then forced to suspend public surveys by the engineers and subscriptions to further projects by the Board. Three road surveys, deemed highly important, were, however, permitted to continue. These were for the road from Staunton to Parkersburg, the road from Cumberland Gap to Giles Court House, and the road from Giles Court House by way of Fayette Court House to Kanawha Falls. A generally successful attempt was also made to keep all improvements companies in which the state had invested from going under. This battle itself, successfully waged in the face of great odds in a time of economic chaos, was quite an accomplishment for the Board for it would allow resumption of work on the uncompleted sections in the mid-1840's when times did improve.

By 1840 the dreams cherished by Virginians since the early days of the republic and Washington's report to Governor Harrison were well on their way to realisation. Several through routes to the Ohio now existed along the lines of the old Indian paths. now upgraded into roads and turnpikes and shortly to become railroads in several cases. Though the old county road system still remained, it had been supplemented by the programs of the Board of Public Works and its technological. and administrative innovations as trade with the West mounted during the first half of the nineteenth century and produced its attendant demand for better transportation. By 1840 also it was becoming apparent to all that the railroad was the idea whose time had come. Increasingly now. the thrust of the Board of Public Works' effort would be toward railroad development. with turnpikes and plank roads coming to be looked upon as feeder lines to the railroads rather than as primary routes themselves. In fact a number of the later turnpikes, such as the one from the Virginia Central Railroad at Gordonsville into the Valley. seem to have been built about 1850 primarily to connect with railroads. In spite of the apparent success of the railroad as the preferred method of transportation, construction continued on the James River and Kanawha Canal and the Chesapeake and Ohio Canal into the 1850's. Nor were the various river navigations abandoned. Some of these continued to operate long after the demise of the James River and Kanawha Canal. The Willis River Navigation in Cumberland County, for instance. continued in use until the late 1890's. transporting produce to the railroad which replaced the canal along the James River. All of these continued to coexist with the railroad until the 1860's when the war came.

The war laid a particularly heavy hand on the railroads since they could so easily be destroyed, although canals too suffered at the hands of the Yankee invader. Roads suffered more by way of neglect than through actual wanton destruction though both sides indulged their incendiary proclivities on the numerous wooden and covered bridges throughout the state. In a sense the war might be viewed as a discriminator, for if railroads, canals, and roads all suffered destruction as well as wear and tear, the last two being viewed as inefficient and archaic in the years after the war could not attract the northern capital necessary to their recovery. The railroads could, and thus swept the field. The turnpikes could not, and most of them soon reverted to county control, while the James River Canal survived a little over a decade, when the damage sustained in the flood of 1877 put it on its way to extinction.

During the post-bellum years the state could do little. Of the engagements during the war, some sixty percent were fought in Virginia and the state had emerged with its economy shattered. Facing a bleak financial situation and racked by political conflict over the funding of its prewar debt, it could do little to aid the internal improvements companies which had been chartered and had bloomed under its sponsorship during the ante-bellum period. The Board of Public Works, much reduced from its former self, tottered along into the twentieth century before it was finally abolished, and roads gradually fell into a state not only worse than that of 1860 but probably worse than that existing before the turnpike era began in the late eighteenth century. What would finally reverse the trend toward decay, and ultimately relegate the railroads to a position not unlike that occupied by highways in 1890, was the booming popularity of the motor car in the early years of this century, and particularly after World War I. Only then would roads recover their old prominence, but this begins to impinge on a story which is more properly that of the present Virginia Department of Highways and Transportation, its origins in the Act of 1906, and its subsequent development into a modern organisation capable of ministering to the varied transportation needs of millions of highly mobile Virginians.

NOTES:

NOTES:

SELECTED BIBLIOGRAPHY

I.   Unpublished Dissertations and Theses

Boughter, Isaac F., "Internal Improvements in Northwestern Virginia", unpublished Ph.D. Dissertation, University of Pittsburgh, 1931.

Harrison, Joseph H., "The Internal Improvement Issue in the Politics of the Union, 1783-1825", unpublished Ph.D. Dissertation, University of Virginia, 1954.

Rice, Philip M., "The Virginia Board of Public Works, 1816-1842", unpublished M.A. Thesis, University of North Carolina, 1947.

Rice, Philip M., "Internal Improvements in Virginia, 1775-1860", unpublished Ph.D. Dissertation, University of North Carolina, 1948.

Roberts, Edward G., "The Roads of Virginia, 1607-1840", unpublished Ph.D. Dissertation, University of Virginia, 1950.

Sarles, Frank B., Jr., "Trade of the Valley of Virginia, 1789-1860", unpublished M.A. Thesis, University of Virginia, n.d.

Tanner, Carol M., "Joseph C. Cabell, 1778-185611, unpublished Ph.D. Dissertation University of Virginia, 1948.

Turner, Charles W., "The Louisa Railroad, 1836-1850", unpublished M.A. Thesis, University of North Carolina, 1940.

Turner, Charles W., "The Virginia Railroads, 1828-1860", unpublished Ph.D. Dissertation, University of Minnesota, 1946.

II.   Printed Public Records

Hening, William Waller, The Statutes at Large. Being a Collection of All the Laws of Virginia 1619-1792, 13 vols, Richmond 1818-1825.

Shepherd, Samuel, The Statutes at Large of Virginia. from October 1792 to December Session 1806. ..Being a Continuation of Hering, 3 vols. Richmond: Samuel Shepherd, 1835-36.

Acts of the General Assembly of Virginia. Richmond: State Printer, 1807-1860.

The Revised Code of the Laws of Virginia, 2 vols. Richmond: Thomas Ritchie, 1819.

The Code of Virginia. Richmond: William F. Ritchie, 1849.

The Code of Virginia, Second Edition. Richmond: Ritchie, Dunnavent and
Company, 1860.

Annual Reports of the Board of Public Works. Richmond: State Printer, 1817-
1860.

Other Heritage Books by the Virginia Genealogical Society
Published with Permission from the Virginia Transportation Research Council
(A Cooperative Organization Sponsored Jointly by the Virginia Department of Transportation and the University of Virginia):

A Brief History of Roads in Virginia, 1607–1840

A Brief History of the Staunton and James River Turnpike [Virginia]

A Guide to the Preparation of County Road Histories

Albemarle County [Virginia] Road Orders, 1725–1816

Albemarle County [Virginia] Road Orders, 1744–1748

Albemarle County [Virginia] Road Orders, 1783–1816

Amelia County [Virginia] Road Orders, 1735–1753

An Index to Roads Shown in the Albemarle County Surveyors Books, 1744–1853

Augusta County [Virginia] Road Orders, 1745–1769

Brunswick County [Virginia] Road Orders, 1732–1746

Early Road Location [Virginia]: Key to Discovering Historic Resources?

Fairfax County [Virginia] Road Orders, 1749–1800

Final Report: Culpeper County [Virginia] Road Orders, 1763–1764

Frederick County, Virginia Road Orders, 1743–1772

Goochland County [Virginia] Road Orders, 1728–1744

Louisa County [Virginia] Road Orders, 1742–1748

Lunenburg County [Virginia] Road Orders, 1746–1764

New Kent and Hanover County [Virginia] Road Orders, 1706–1743

Orange County [Virginia] Road Orders, 1734–1749

Spotsylvania County [Virginia] Road Orders, 1722–1734

The Route of the Three Notch'd Road [Virginia]: A Preliminary Report

Other Heritage Books by the Virginia Genealogical Society:

Magazine of Virginia Genealogy:

Volume 22, Numbers 1–4, 1984

Volume 23, Numbers 1–4, 1985

Volume 24, Numbers 1–3 plus index, 1986

Volume 25, Numbers 1–4, 1987

Volume 26, Numbers 1–4, 1988

Volume 27, Numbers 1–4, 1989

Volume 28, Numbers 1–4, 1990

Volume 29, Numbers 1–3 plus index, 1991

Volume 30, Numbers 1–4, 1992

Volume 31, Numbers 1, 2, 4 plus index, 1993

Volume 32, Numbers 1–4, 1994

Volume 33, Numbers 1–4, 1995

Volume 34, Numbers 1–4, 1996

Volume 35, Numbers 1–4, 1997

Volume 36, Numbers 1–4, 1998

Volume 37, Numbers 1–4, 1999

Volume 38, Numbers 1–4, 2000

Volume 39, Numbers 1–4, 2001

Volume 40, Numbers 1–4, 2002

www.ingramcontent.com/pod-product-compliance
Lightning Source LLC
Chambersburg PA
CBHW081242090426
42738CB00016B/3384

9 780788 436741